Yellow Umbrella Books are published by Red Brick Learning
7825 Telegraph Road, Bloomington, Minnesota 55438
http://www.redbricklearning.com

Library of Congress Cataloging-in-Publication Data
Trumbauer, Lisa, 1963-
   [What makes ten. Spanish & English]
   What makes ten/by Lisa Trumbauer = ¿Qué hace diez?/por Lisa Trumbauer.
   p. cm.
   Summary: "Simple text and photos introduce the concept that objects can be
grouped in different ways to equal ten"—Provided by publisher.
   Includes index.
   ISBN-13: 978-0-7368-6016-1 (hardcover)
   ISBN-10: 0-7368-6016-9 (hardcover)
   1. Addition—Juvenile literature. 2. Arithmetic—Juvenile literature. I. Title: ¿Qué hace
diez? II. Title.
QA115.T77718 2006
513.2'11—dc22   2005025847

Written by Lisa Trumbauer
Developed by Raindrop Publishing

Editorial Director: Mary Lindeen
Editor: Jennifer VanVoorst
Photo Researcher: Wanda Winch
Adapted Translations: Gloria Ramos
Spanish Language Consultants: Jesús Cervantes, Anita Constantino
Conversion Assistants: Jenny Marks, Laura Manthe

Photo Credits
Cover: DigitalVision; Title Page: Photo 24/Brand X Pictures 24; Page 4: Deirdre
Barton/Capstone Press; Page 6: Deirdre Barton/Capstone Press; Page 8: G. K. & Vikki Hart/
PhotoDisc; Page 10: Photo courtesy of Jenny Peacocke; Page 12: DigitalVision; Page 14:
DigitalVision; Page 16: Steve Mason/PhotoDisc

1 2 3 4 5 6 11 10 09 08 07 06

What Makes Ten
by Lisa Trumbauer

¿Qué hace diez?
por Lisa Trumbauer

Yellow
Umbrella
Books
for early readers

One flower here

Una flor aquí

plus nine flowers here –
that makes ten.

más nueve flores aquí,
hacen diez.

Three puppies here

Tres perritos aquí

plus seven puppies here –
that makes ten.

más siete perritos aquí,
hacen diez.

Five children here

Cinco niños aquí

plus five children here –
that makes ten.

más cinco niños aquí,
hacen diez.

What makes ten here?

¿Qué hace diez aquí?

Index

Índice